Abdelhafid Guediri
Mourad Guediri
Abdelkarim Guediri

Fundamental Electrical Engineering II

Abdelhafid Guediri
Mourad Guediri
Abdelkarim Guediri

Fundamental Electrical Engineering II

Courses and exercises

ScienciaScripts

Imprint

Cover image: www.ingimage.com

This book is a translation from the original published under ISBN 978-620-6-71479-8.

Publisher:
Sciencia Scripts
is a trademark of
Dodo Books Indian Ocean Ltd. and OmniScriptum S.R.L publishing group

120 High Road, East Finchley, London, N2 9ED, United Kingdom
Str. Armeneasca 28/1, office 1, Chisinau MD-2012, Republic of Moldova, Europe
Printed at: see last page
ISBN: 978-620-8-09420-1

Contents

Authors

Dr. Guediri Abdelhafid

Dr. Guediri Mourad

Prof. Guediri Abdelkarim

INTRODUCTION

A special-purpose **machine** is a production machine designed to meet the specific needs of an industrial company, as opposed to general-purpose machines such as machine tools. It differs from other production machines by its technical originality and, as its name suggests, by the fact that it is not generic. It is often manufactured to meet unique automation needs, and is most often used for manufacturing, assembly, testing or packaging.

2. The different types of special machines range from workstations to fully automated production lines:

The design of a **special machine** requires the expertise of a design office with knowledge in various industrial fields: **Automation**, in particular the use of industrial PLCs **Mechanical engineering**, with design and manufacturing problems **CAD** for the drawing of specific parts and for modelling the whole machine with all its components **Pneumatics**, with the use of cylinders to manage movement Know-how of parts or tools

Industrial vision in 2D, 3D, thermal or X-ray for position control or adjustment, for example **Industrial robotics** with the programming of 3 to 6-axis robots **Electricity** for the production of electrical diagrams and the wiring of all electrical components Special-purpose machines can be found in all branches of industry and for a wide variety of needs, such as..: **Sectors of activity and examples of achievements Aeronautics:** automatic cell for assembling and dismantling machining inserts **Sanitary equipment: fully** automated line for assembling, testing and packaging taps **Automotive: semi-automatic assembly** of silent blocks and soundproofing for Renault engines Pharmaceuticals: fully automated station for assembling valves for the medical sector Animal health industry: **cutting** and packaging of animal feed Electronics: circuit breaker winding machines

CHAPTER 1

Universal motor

A **universal motor** is an electric motor operating on the same principle as a series-excited **DC machine**: the rotor is connected in series with the field winding, so that the rotor and field currents are always in the same direction. The torque of this machine is independent of the direction of current flow and is proportional to the square of its intensity. It can therefore be supplied with either **direct** or **alternating current**, hence its name.

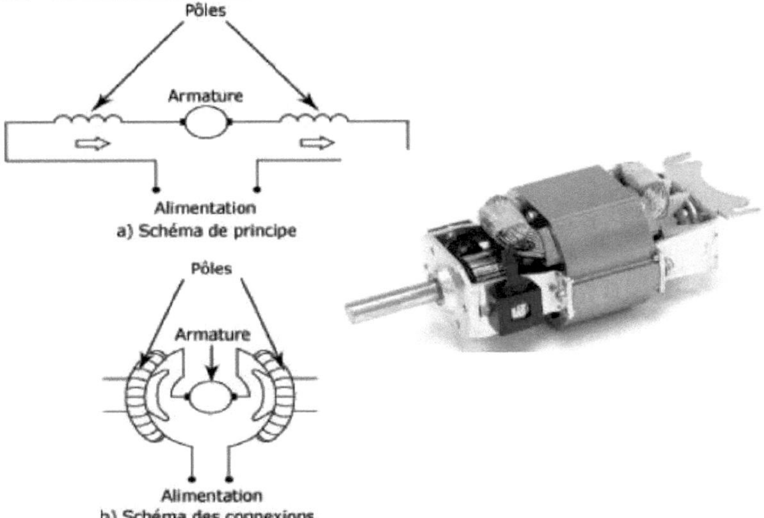

Figure1: universal motor

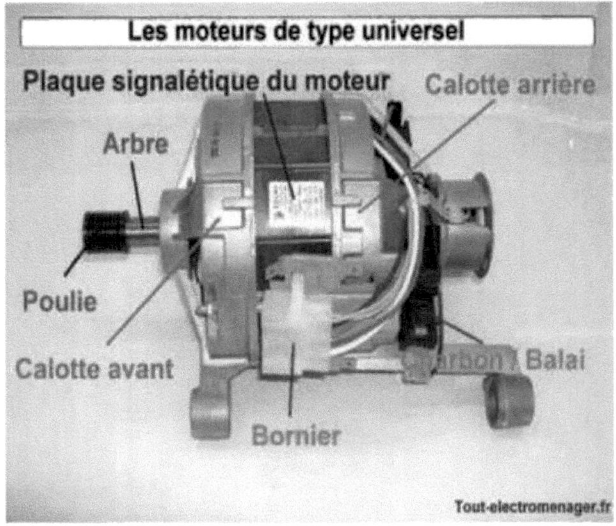

Figure2: universal motor

1) Description :

In order to limit the eddy currents that systematically occur in all solid metal areas subjected to alternating magnetic fields, its stator and rotor are laminated. Generally speaking, the efficiency of this type of machine is poor (but their manufacturing cost is reduced), their torque is low, but their rotation speed is high. When they are used in devices requiring high torque, they are combined with a mechanical gearbox.

This is particularly the case for low-power household appliances and hand-held power tools (up to around 1.2 kW) and many domestic applications.

These motors are also widely used in hoovers. In this case, the turbine is in direct contact with the motor and the air sucked in circulates in the air gap, cooling the motor.

The speed of rotation of these motors is proportional to the value of the supply voltage. Under AC conditions, it can be easily adjusted by an inexpensive device such as a phase-angle dimmer (the same type of dimmer used to adjust the light intensity of luminaires).

2) Disadvantages:

- Poor yield (20% to 40%).
- Wear of the brushes supplying the rotor, as in any DC machine.

Successive breaks in contact, inherent in the operation of the brush-collector assembly, generate interference in the power supply circuit and electromagnetic and radio interference for many other devices: televisions, radios, telephones, etc.

3) Advantage :

- Very low manufacturing costs.
- Easy speed variation.

Note:

Given the poor efficiency of this type of motor, we must not confuse the absorbed power (that which is delivered to the motor by the electrical supply system) with the useful power (that which is transmitted by the motor to the equipment it drives). This is particularly true when talking about the power of a hoover.

CHAPTER 2

Linear motor

Linear motors are a special category of brushless synchronous servomotors. They operate on the same principle as torque motors, but are open and run flat. The electromagnetic interaction between an assembly of coils (primary assembly) and a track of permanent magnets (secondary assembly) transforms electrical energy into linear mechanical energy with great efficiency. The primary assembly is also usually referred to as the motor, moving part or carriage; the secondary assembly is also referred to as the magnet track or v Since linear motors are designed to produce high forces at low or zero speed, sizing is based not on power but on force, unlike drives.
traditional.

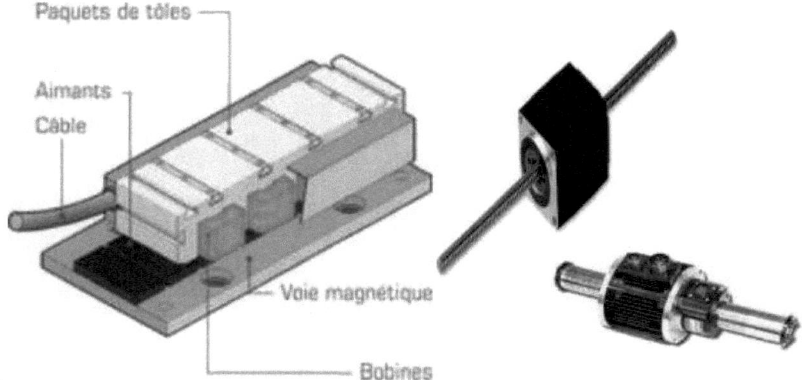

Figure 3 : Linear motor

As the moving part of a linear motor is directly attached to the machine load, this saves space, simplifies machine design, eliminates mechanical play and eliminates sources of failure caused by belts, ball screws and other transmission elements. Finally, the bandwidth and rigidity of a positioning system are much improved, giving greater positional accuracy and precision over an unlimited stroke at high speeds. As linear motors do not include cages, bearings or position encoders, the machine builder is free to select the additional components of his choice to best meet the requirements of the application.

1) Applications :

As rotary motors, linear motors are entered in the following fields:

- Transport (trains)
- Lifts.
- Tie ports.

2) The advantages of the linear engine :

- High speed.
- High-precision microcontroller controlled via the reverse power supply.
- Rapid response.
- Operation.
- Durability where no mechanical connections are used to obtain a linear movement (such as a case obtained from a rotary engine).

3(The disadvantages of the linear engine:

- cost (high cost of the magnets used and high cost of the linear encoders used for the reverse power supply).
- the controller used is the most sophisticated of those used in rotary motors.
- high temperature due to its own structure.

CHAPTER 3

Asynchronous tachogenerator

1) Definition :

Tachometric, adj. Generator, tachometric -tric. Instrument formed by an electric generator coupled to a rotating member and connected to an electric measuring system (from Encyclop. Sc. Techno. t. 10 1973, p. 262a). The essence of High Fidelity Radio equipment lies in one concept: accuracy. Mechanical accuracy: the capstan is driven by a quartz-controlled direct-current motor, and the tape deflection is controlled by a tachometric generator.

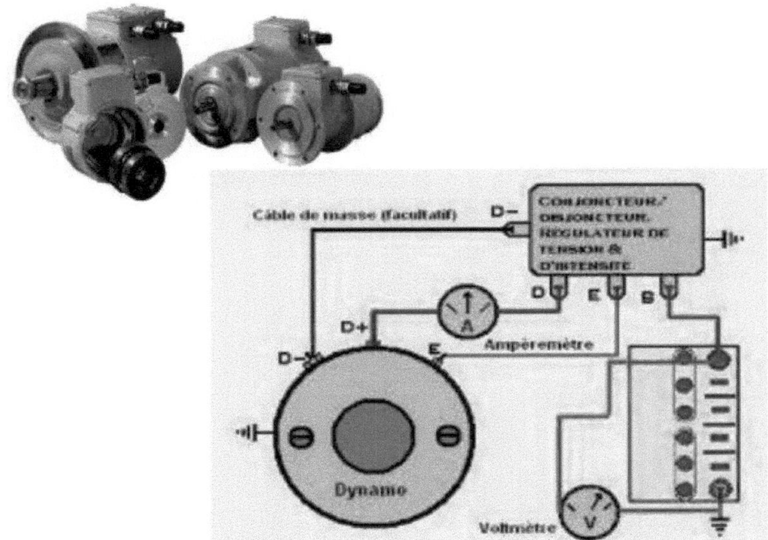

Figure4 : Tachometric generator

2) How it works:

This transducer works like a DC electric motor (see appendix 1 on electric motors), but in the opposite direction: the voltage it delivers is proportional to its speed of rotation. This can be demonstrated by looking at the DC motor equations, for negligible resistance and inductance. This sensor therefore delivers analogue information: a voltage proportional to the measured speed of rotation.

3)Scope of application :

It delivers a voltage proportional to its speed of rotation. Its main field of application is speed regulation of an electric motor J

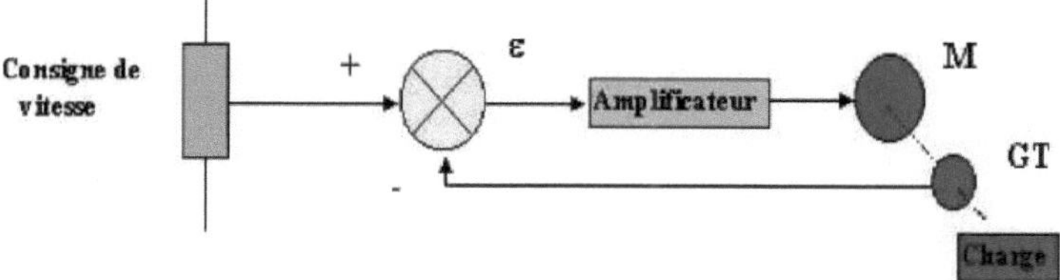

Figure 5: speed control of an electric motor with GT

4) Essential characteristics of tachometry :

- maximum speed of rotation (in revolutions per minute),
- e.m.f. constant (in volts at 1000 rpm or in v/rev/mn),
- linearite (in %),
- crete to crete undulation (in %),
- maximum current

CHAPTER 4

Stepper motors

1) Definition :

A stepping **motor** is a DC motor that moves forward in a single step when the direction of the current in one of the coils changes.

2) Operating principle :

Stepper motors are reluctance micro-motors (also known as variable reluctance motors), which operate by attracting a polar mass (the rotor) by a magnetic field.

- Each phase (1 - 2 - 3) of the stator receives an electrical impulse in turn.
- Conventionally positive or negative depending on the direction of the current in the coil.
- The pulses arrive in a predetermined order of distribution and at a predetermined, adjustable frequency.
- Each of these pulses corresponds to an angular displacement called a "step".

A "step" is a unit of angular displacement, generally 1.8 degrees (the motor will then have 200 steps) 1.8 x 200 = 360

3) The different types of step-by-step :

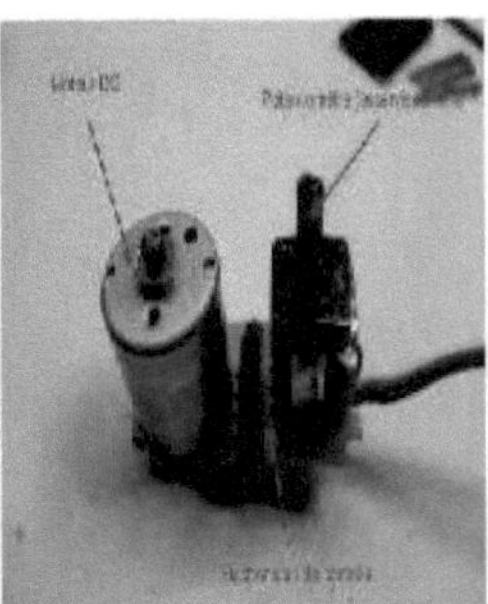

Figure6: stepper motors

- Three categories of engine:

> **With** the same electrical characteristics, such a motor is less powerful, but faster than permanent magnet motors. Probably the oldest.

> **With permanent magnets:** you can feel the steps. These are low-cost motors with medium resolution (up to 100 steps/revolution).

> **Hybrids**: These engines combine the 2 previous technologies, and are more expensive. Their advantage lies in better torque, higher speed, and a resolution of 100 to 400 steps/revolution.

> The most common motors are permanent magnet and hybrid.

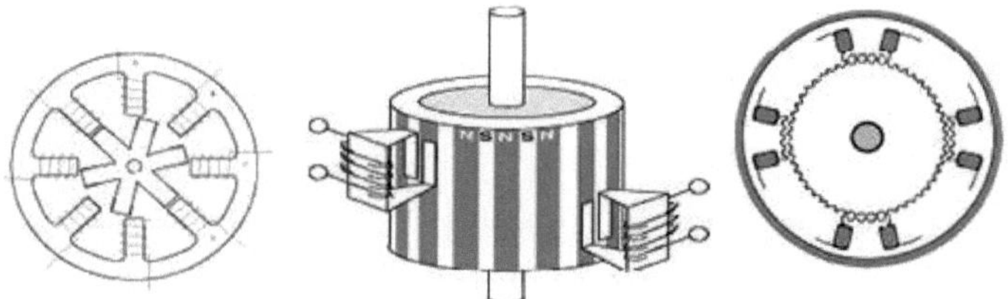

Figure7: Hybrid Variable Reluctance Permanent Magnets

4) Comparison of step-by-step types:

- Bipolar:

- Higher available power for identical mechanical characteristics.

- Single-pole:

- The cheapest!
- Easier to implement. This was especially true before the arrival of the

5) Permanent magnet motors with bipolar power supply

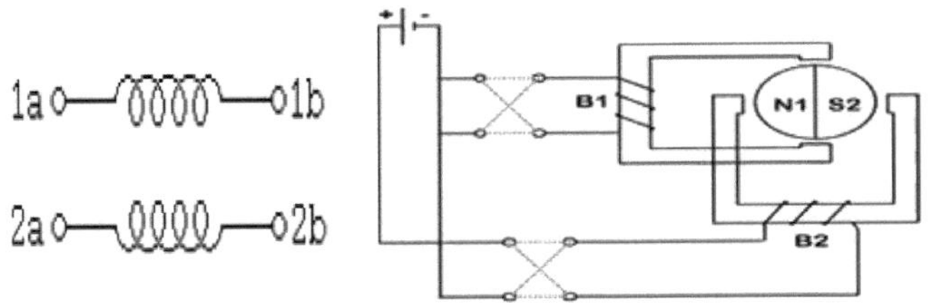

Figure8: type 1 stepper motors

A bipolar motor: is a motor with two stator phases B1 and B2 without a centre point and two switches.
The direction of flow is reversed by reversing the direction of the current.

6) Permanent magnet motors and unipolar power supply

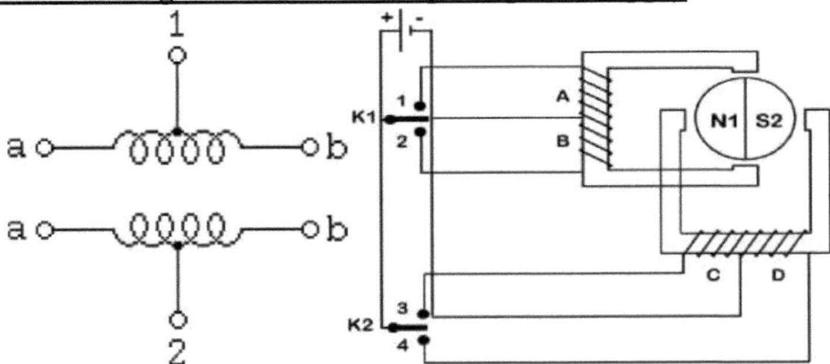

Figure9: type 2 stepper motor

> **The benefits:**
> Constant rotation for each command (accuracy better than 5% of a step).
> A couple at a standstill.
> Control of position, speed and synchronisation of several motors (no need for contra-reaction)
> Brushless motor.
> **The drawbacks:**
> More difficult to operate than a DC motor.
> Relatively low speed and torque.

- Rapidly decreasing torque as speed increases.
- Mechanical resonance.

CHAPTER 5

General concept of the synchronous machine

1) Introduction

The term synchronous machine covers all machines where the speed of rotation of the output shaft is equal to the speed of rotation of the rotating field.

To achieve this, the rotor magnetic field is generated either by magnets or by an excitation circuit. The position of the rotor magnetic field is then fixed relative to the rotor, which in normal operation imposes an identical speed of rotation between the rotor and the rotating stator field.

This family of machines actually comprises several sub-families, ranging from alternators of several hundred megawatts to stepper motors and motors of just a few watts.

Nevertheless, the structure of all these machines is relatively similar. The stator is generally made up of three three-phase windings, so that the electromotive forces generated by the rotation of the rotor field are sinusoidal or trapezoidal.

The stators, particularly at high power, are identical to those of an asynchronous machine.

There are three main types of rotor, whose role is to generate the rotor induction field:

- smooth pole coil rotors;
- coil rotors with salient poles;
- magnet rotors.

2) General information:

The synchronous machine is a reversible electro-mechanical conversion machine. It can be found in many energy conversion devices, both in:

- production of electrical energy from mechanical energy, where it is called a **synchronous generator** when the speed is variable (e.g. wind turbine) or an **alternator** when the speed is fixed (e.g. thermal power station),
- production of mechanical energy from electrical energy, where it is called a **synchronous motor** (e.g. TGV traction chain).

With the development of power electronics, synchronous motors are increasingly replacing DC motors. The absence of a brush-collector device, whose function is transferred to the power electronics, eliminates the problem of maintenance and speed limits.

Example of the inside of a synchronous machine

As with all electro-mechanical converters, it is the interaction of 2 magnetic fields that produces mechanical or electrical energy.

3) **Constitution of the synchronous machine:**

3-1- General provisions:

The synchronous machine consists of a stator (fixed) and a rotor (mobile).

- The **stator** is of the same nature as that of the asynchronous machine. It consists of a stack of notched ferromagnetic plates supporting a three-phase winding whose angular geometry is defined by the number of pairs of poles. It is in this winding that the currents flow.
- The **rotor**, also known as the pole wheel in the case of the alternator, has a magnetic structure with p pairs of poles. This magnetic structure can be produced using magnets or coils supplied with direct current.

Magnet rotor

Rotor with smooth poles

Rotor à pôles saillants

3-2- Number of poles in a machine:

a) Definition:

The number of pole pairs p corresponds to the number of North-South magnetic patterns around the circumference of the rotor.

b) Symbols:

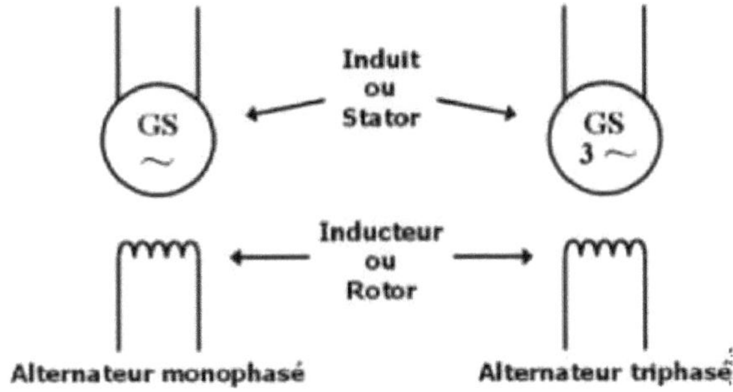

3-3- <u>Generator operating principle:</u>

a) Creation of e.m.f.s.

When the North poles of the rotor face the phase 1 coils of the stator,

i

Later, the South poles face these coils and the flux through this phase is then minimum. And so on.

- The flow through a phase is alternating of frequency f.

If N is the rotor speed in revolutions per second, the period **T** of the flow through

a phase is equal to the duration of $\frac{1}{p}$ turn. Hence :

$$T = \frac{1}{p.N} \qquad f = p.N \qquad \omega = 2\pi.pN$$

The e.m.f. induced in this phase $e = -n\frac{d\varphi}{dt}$ is alternating and of the same frequency.

The windings of phases 2 and 3 of the stator behave identically.

2тr 4ir

to that of phase 1, but shifted by $\frac{2\pi}{3p}$ et $\frac{4\pi}{3p}$ The flow is at its maximum $\frac{4\pi}{3p} \times \Omega$ ou $\frac{2\pi}{3p}$.

later. The difference between the induced e.m.f.s is $\frac{T}{3}$.

The result is an equilibrium three-phase system with an e.m.f. of frequency f.

b) Creating electromagnetic torque:

When the 3 phases are connected to a receiver, they produce a three-phase system of currents. They thus create **a** magnetic field of angular velocity:

$$\omega_s = \frac{\omega}{p} = 2\pi N = \Omega$$

c) Principle of the synchronous machine in generator mode:

The stator rotating field rotates at the same speed as the rotor.

- Remarks :

- for an alternator: the electromagnetic torque is a braking torque.
- for a motor, this torque is a motor.
- we have to be careful in the power balances to know what ***a*** o

power : $C_{em}.\Omega$

- Electromotive force:

e.m.f. induced by a coil:

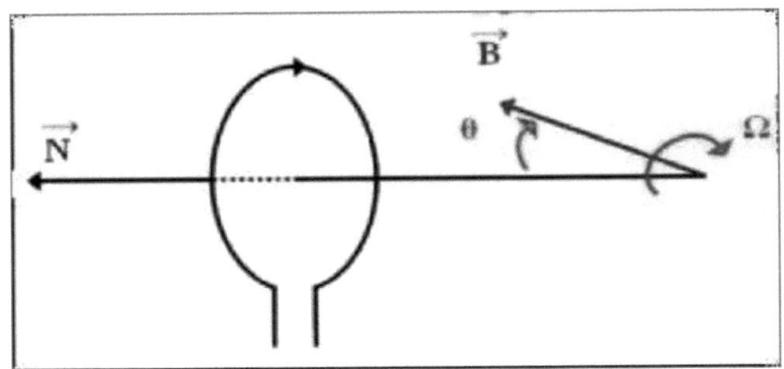

Let's consider a turn of area S whose normal makes an angle θ with the magnetic field created by a magnet rotating at a constant angular velocity $\Omega = 2\Pi n$. Let's choose as the origin of time the instant at which the angle $\theta=0$: the flux through the turn is then maximum: : $\hat{\Phi}=B.S$

During the time $\Delta t=t-0=t$, the vector £ rotates by an angle $\theta=\Omega.t$. The flux at time t is : $\varphi = B.S.\cos(\theta) = \hat{\Phi}.\cos(\Omega.t)$.

As the field rotates, the flux varies and an induced e.m.f. appears at the terminals.

of the spiral : $e = -\frac{d\varphi}{dt} = \Omega.\hat{\Phi}.\sin(\Omega.t)$.

Its rms value is: $E1 = \frac{E1}{\sqrt{2}} = \frac{\Omega.\Phi}{\sqrt{2}} = \frac{2\Pi n.\Phi}{\sqrt{2}} = 4,44n\hat{\Phi}$

c.1) Case of the single-phase alternator:

The pole wheel develops 2p poles of the field strength. The period of the e.m.f. induced in a turn is divided by p. Everything happens as if the magnet were rotating at speed .-". Hence the rms value of the e.m.f. induced in a turn is: #=4Mp.п.Ф

As the stator has N turns, the total e.m.f. across the stator winding is :

$$E = 4,44.N.p.n.\hat{\Phi}$$

c.2) Case of the three-phase alternator:

The following relationship will suffice:

$$E_j = K.p.n.N.\hat{\Phi}$$

K : *Kapp* coefficient (shortening coefficient, diametral pitch, etc.)

CHAPTER 6

Operation of the Synchronous Machine

1. Characteristics of an isolated alternator:

At no load, the stator couples like a star, the armature discharges no current. The rotor is driven at a constant nominal speed n. We note :

-This characteristic is similar to that of an MCC (Direct Current Machine) and the magnetisation curve of the machine's magnetic circuit is clearly visible:

- The useful zone is in the vicinity of the saturation bend
- Sometimes a hysteresis phenomenon doubles the characteristic

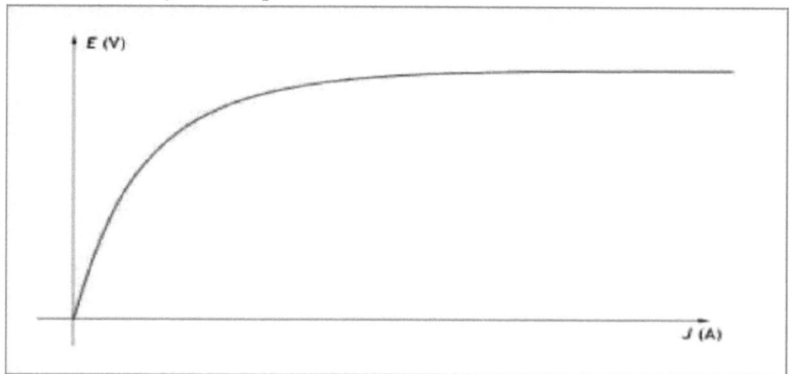

2. Synchronous reactance diagram:

The simplest way to account for the on-load operation of an alternating generator is to treat it as an e.m.f. source. E with an internal impedance R+jX.

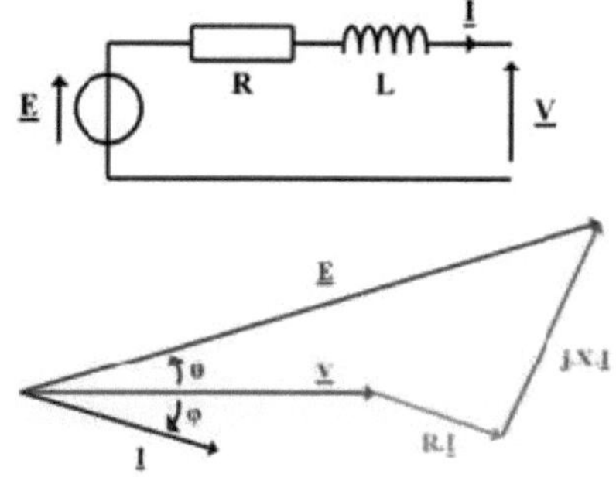

E: vacuum e.m.f.

V: voltage at the homes of a machine winding

/? : winding resistance

X: synchronous reactance (lei X = L)

θ : internal corner

This equivalent diagram and the vector diagram allow us to go from the regime at terminals *a)* to the e.m.f. в that the field flux (rotor) must create.

The law of meshes gives us :

$$\underline{E} = \underline{V} + (R + jX)\underline{I}$$

Comments:

- The regime at the load terminals depends on the load being supplied. In the previous figure, the load is inductive because it absorbs a phase current behind the voltage.
- я 'must be small because з.я gives the joule losses in the armature. On the contrary, *яе* is very high because it accounts for all the flux created by the armature.
- The reactance X has no physical significance unless we neglect the saturation of the magnetic circuit. It is called synchronous reactance to distinguish it from reactances that occur in the transient regime.

Note on the specification of an alternator:

An alternator is characterised by :

- its frequency
- its compound voltage
- its apparent nominative power
- and a power factor value $\cos\varphi$

Now $\cos\varphi$ does not depend on the machine but on the load it supplies.

The rated current In is given by the apparent power.

The power factor indirectly indicates the excitation current, i.e. the value of $\cos\varphi$ for which the flow of In requires the rated excitation current.

3. Self-piloting synchronous machine:

The synchronous machine can only operate at constant speed. Its mechanical characteristics can be summarised as follows:

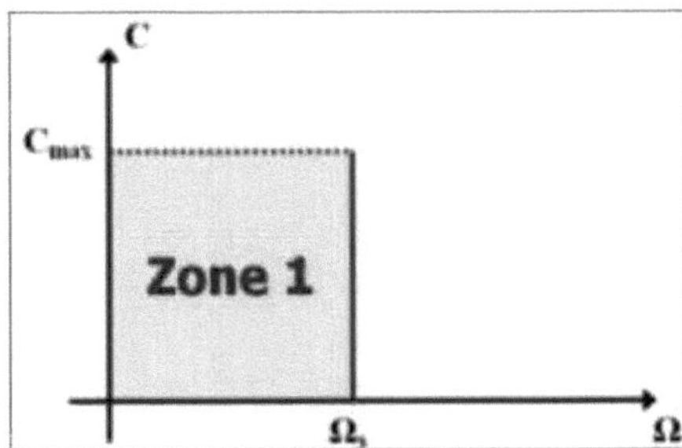

-As the speed of rotation is linked to the supply frequency, the characteristic can

be summarised as a straight line segment.
However, using a variable frequency drive, it is possible to operate in zone 1. The problem in zone 1 is to keep control of the torque. To do this, we loop the MS. The machine is said to be **self-piloting**.

a) Principle of Operation of the Self-Piloted Synchronous Machine:

- A current sensor is used to generate switching commands for an inverter which supplies the stator with voltage v and frequency f.
- A position sensor mechanically fixed to the rotor measures angle ϑ , i.e. the angular position of the rotor relative to the stator field. After multiplication, this makes it possible to control the torque, since the latter is proportional to the stator field. $J.\sin(\vartheta)$.

This self-controlled machine is equivalent to a MCC, since the assembly (collector + inverter) plays the role of the brush-collector assembly. As a result, we can use the same equations. (E proportional to n , and Cm to I (current in one phase)).

b) Power balance:

- **Collected or absorbed power:**

An alternator recycles the mechanical power P_{mec} supplied by the drive system (turbine, diesel engine) and the electrical power P_e supplied by the rotor.

$$P_{mec} = T_m . \Omega$$

or T_m is the drive torque moment that can be measured:

$$P_e = U_e . J = R_e . J^2 = \frac{U_e^2}{R_e}$$

where R_e, is the resistance of the pole wheel winding
The absorbed power is then written :

$$P_a = T_m . \Omega + U_e . J$$

- **Power and efficiency:**

The alternator is a three-phase voltage source supplying a three-phase load with a power factor of $\cos\varphi$. . The useful power is written as done :

$$P_{active} = \sqrt{3} . U . I . \cos\varphi$$

with U line voltage in (V), and I rms line current in (V).

(A)

As with all machines, efficiency is defined by :

P_{rn} ecanique ri -

P_e lect

Orders of magnitude: Power station alternators have efficiencies of around 98%.

- **Losses:**
- **joules in the inductor**: equal to the power it receives:

$$P_e = U_e . J = R_e . J^2 = \frac{U_e^2}{R_e}$$

- **joules in the armature**: As with the MAS, if R is the resistance measured between two stator terminals already coupled :

$$P_{js} = \frac{3}{2} R . I^2$$

- **constants** : They correspond to the sum of the mechanical and magnetic losses (field and armature) and do not depend on the load. They depend on the frequency and voltage.

4. MS coupled to a high-power network:

a) Principle and reversibility:

The alternator coupled to a distribution network operates at no load. If the drive motor is uncoupled from the rotor, the rotor continues to turn synchronously thanks to the stator field.

The machine now becomes a motor, capable of transforming electrical energy from the grid into mechanical energy.

b) Characteristics and stability:

- **Motor speed :**

The speed of the synchronous motor is imposed by the mains frequency because /=P^n . It is therefore **independent of the motor load**.

- **Single-phase scheme and torque :**

Since the single-phase alternator diagram was constructed on the MS principle, it remains valid for synchronous motor operation. The receiver convention is sometimes used to show that the alternator is operating as a motor.

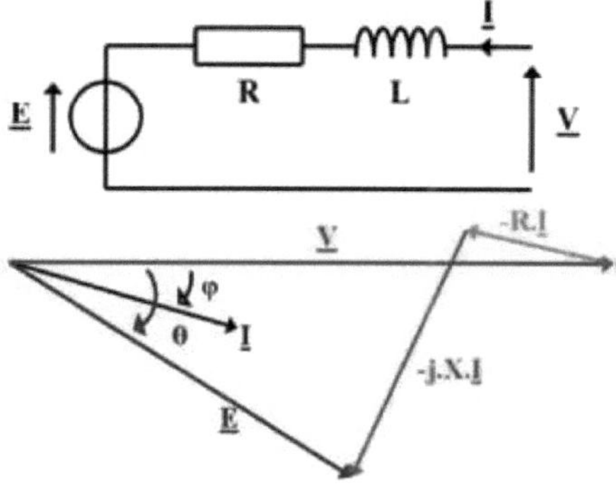

Hence the expression of the useful torque and the electromagnetic torque:

$$T_{em} = \frac{\sqrt{3}U.I.\cos\varphi}{\Omega}$$

And

$$T_u = \eta\frac{\sqrt{3}U.I.\cos\varphi}{\Omega}$$

5. Advantages and disadvantages of the Synchronous Machine

5.1. Uses of the Synchronous Machine:

a) Use of a synchronous machine as an alternator

Almost all the electricity generated in France comes from synchronous alternators. These very high-power alternators (up to 1500 MVA) are essentially different from conventional synchronous machines:

- by their geometry: increasing the power of alternators necessarily means increasing their size. In order to reduce the problems associated with normal acceleration at the periphery of the rotor, manufacturers limit the radius of the machines, which leads to an increase in length.
- by their excitation system
- by their cooling

b) Excitation of high-power alternators:

The excitation powers of high-power alternators are such (several megawatts) that it is interesting to use the mechanical power available on the shaft to supply the excitation current. In this case, an excitation system mounted on the same shaft as the alternator rotor is used. In addition, it is then possible to eliminate the sliding contacts necessary to

feeding the excitement :

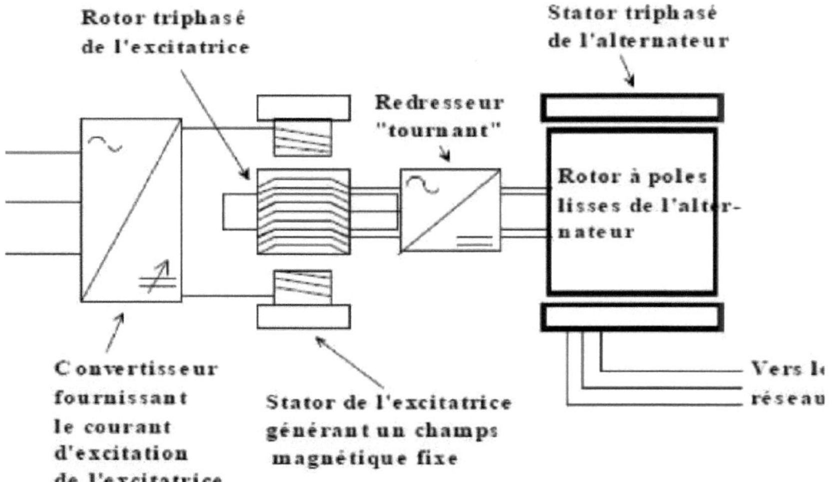

The exciter is in fact an inverse alternator where the excitation circuit is located on the stator. The rotor has a three-phase winding system whose currents are rectified to supply the alternator field.

6. Alternator cooling:

Even if the efficiency of alternators is excellent (close to 99% for a 1000MW alternator), the power dissipated in the form of joule losses is enormous (close to 1MW for a 1000MW alternator) and this in a restricted volume. It is therefore necessary to install heat removal systems based on the use of heat transfer fluids circulating in the stator, rotor and stator conductors. The cooling diagram for a 300MW alternator is shown below:

- **Use of a synchronous machine as a synchronous compensator.** A synchronous compensator is a synchronous machine running at no load whose sole function is to consume or supply reactive power to the grid. By adjusting the excitation current, it is possible to supply reactive energy (if the machine is overexcited) or to consume energy (if the machine is underexcited).

Such machines are used in particular to supply reactive energy when the network is under load, and to absorb the reactive energy generated by the lines when consumption is low.

- **Advantages and disadvantages of the synchronous machine:**

The synchronous machine with permanent magnets on the surface seems to be the best choice for the wheel motor. These machines do indeed have significant advantages:

- High torque/mass and power/mass ratios.
- Very good performance.
- Less wear and tear and lower maintenance costs (no brushes or carbon

brushes).
However, they do have certain drawbacks:

- High cost (due to the price of magnets).
- Problems with magnet temperature resistance (250°C for samarium-cobalt)
- Risk of irreversible demagnetisation of the magnets due to armature reaction.
- Difficult to deflux and complex control electronics (position sensor required).
- Impossibility of regulating 1 excitation.
- To reach high speeds, it is necessary to increase the stator current in order to demagnetise the machine. This will inevitably lead to an increase in stator losses due to the Joule effect.
- The fact that this flow is not regulated means that it cannot be controlled flexibly over a very wide speed range.

7. Conclusion

This work is a presentation on permanent magnet synchronous machines, which are generally three-phase machines. The rotor, often called "pole wheel", is supplied by a direct current source or equipped with permanent magnets. They play a very important role in the industrial field because they are :
High torque/mass and power/mass ratios.
Very good performance.
Less wear and tear and lower maintenance costs

But they also have certain disadvantages, such as :
High cost. Problem with magnet temperature resistance. control (position sensor required).
Impossibility of regulating 1 excitation.
To reach high speeds, it is necessary to increase the stator current in order to demagnetise the machine.

CHAPTER 7

stepper motors with variable reluctance:

1. History: the first variable reluctance stepper motors were used by the British Navy in the 1920s to move the direction indicators on torpedo launchers and cannons. In the 1930s, engineer Marius Lavet discovered a particular type of magnetic stepper motor, now known as the Lavet motor, which enabled this device to be developed in the field of hodology thanks to its miniaturisation and low cost.

The classic stepper motor appeared in the 1940s, but it was the advent of digital electronics in the 1960s that led to its development.

2. Introduction:

The stepper motor is an electromechanical converter that transforms a pulsed electrical signal into a mechanical displacement (angular or linear). Its basic structure consists of two mechanically separate parts, the Stator and the Rotor. Electromagnetic interaction between these two parts ensures rotation **3. Definition:**

The stepper motor is an electromechanical converter designed to transform the electrical signal (pulse or) into mechanical displacement (angular or linear).

From an electrotechnical point of view, the classic motor resembles the synchronous machine, in which the stator (usually with salient poles) carries the drive windings and the rotor (almost always with salient poles) is either fitted with permanent magnets (a so-called polarised or active structure), or consists of a slotted ferromagnetic part (a so-called reluctance or passive structure).

Between the motor and its power supply, there are three essential elements: -a calculation unit, which prepares the control impulses.

- a PWM modulator, which generates the commands for the electronic switching contactors.
- switching electronics (power), which, from a power supply, transfers the energy to the appropriate motor windings

Figure 3. "Two phase on" stepping sequence for two phase motor.

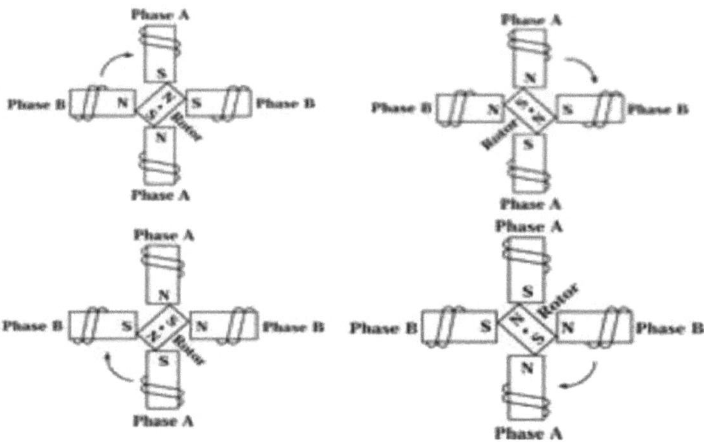

4. Operating principle

- The motor is energised via the stator coils.
- The number of steps depends:
- the number of phases (coil group);
- the number of rotor and stator poles;
- of the motor phase switching sequence.

5. The basic principle:

5.1.The one-step answer:

- When the motor moves forward one step, the rotor response is that which the system would have at one step.
- This oscillatory response can lead to resonance phenomena.

5.2.The sound of the engine :

- At certain speeds, resonance causes the motor to lose steps.
- The resonance speed depends on the load and the motor.

5.3.To dampen the resonance :

- The use of coupling elements with a certain degree of elasticity introduces damping to reduce resonance.

6. Types of stepper motor:

Various stepper motor technologies are available:

- Variable reluctance motor;
- Permanent magnet motor;
- Hybrid engines.

6.1.Variable reluctance motor :

- The stator has a number of teeth with a winding.

- The rotor (made of magnetic material) has a different number of teeth, but no windings.
- The rotor is positioned so that the reluctance of the magnetic circuit is minimal.
- Sequence for a round 1-2-3-1-2-3-1-2- 3-1-2-3-1.
- 12 steps per revolution or 30° per step.

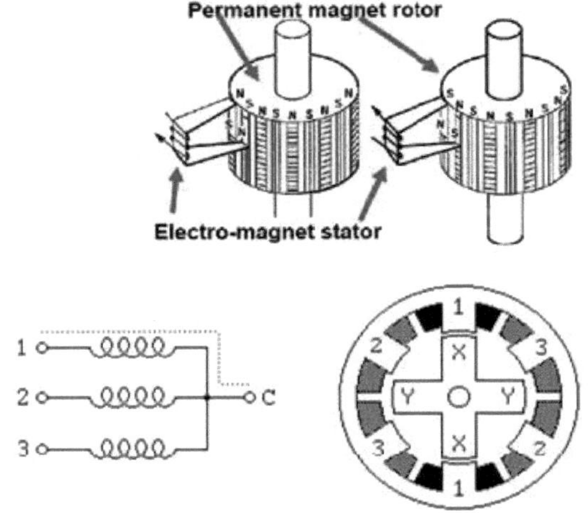

6.2.Permanent magnet motor

- The stator has a number of teeth with a winding.
- The rotor is a magnet that will align with the poles of the stator that are powered.

6.3.Hybrid engines

- Combination of variable reluctance motor and permanent magnet motor.
- Available in two models:
- unipolar and bipolar.
- Unipolar sequence: la - 2a - lb - 2b - la (120°)
- Bipolar sequence: 1$ -2$ -1| - *21* -1$ (120°)

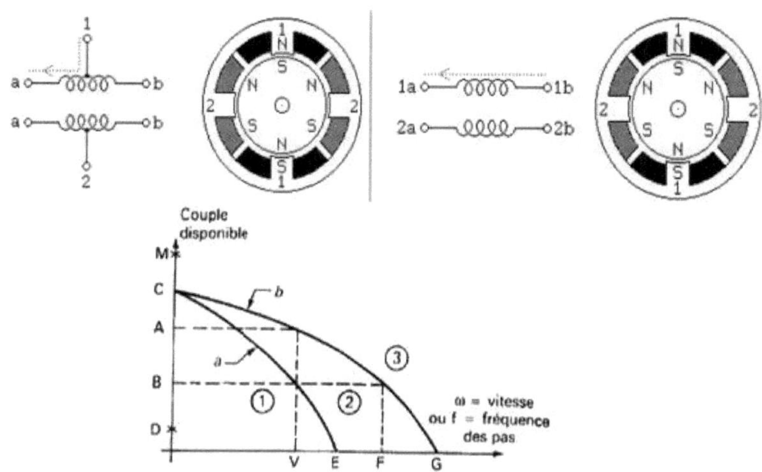

7. Features:

- Torque behaviour as a function of pulse speed:

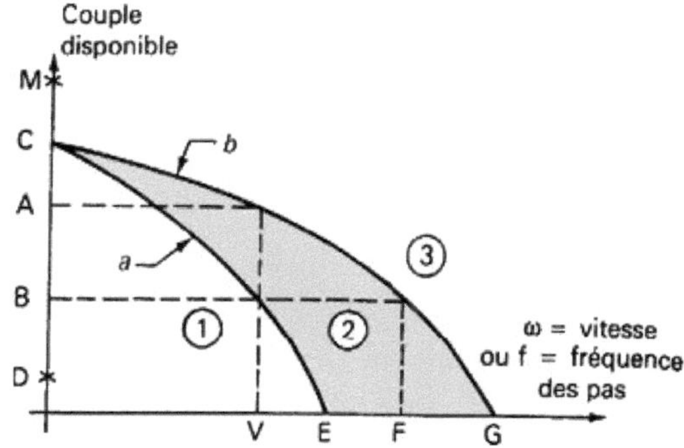

- Start-stop zone:

- B: Maximum starting torque at speed V;
- E: Maximum no-load starting speed;
- M: Holding torque;

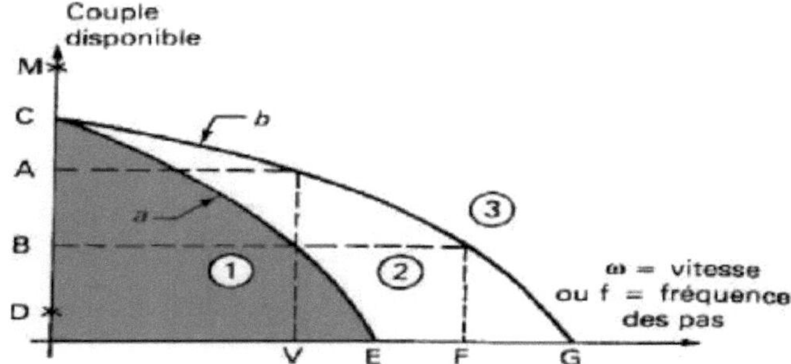

- D: Detent torque (non-powered).

- Training area :

- A: Maximum drive torque at speed V;
- F: Maximum drive speed at torque B;
- G: Maximum unloaded drive speed;
- C: Maximum dynamic torque.

- Characteristics

- Size:

- External motor diameter in 1/10 inch.
- Gives an idea of the engine's power.

- Power:

- Apparent power: product of torque and speed.
- Can reach 4 to 5 kW.

8. Use of stepper motors:

- **Positioning**:

- X-Y control of drawing and electro-erosion machine tables
- Rotary drive for machine-tool circular tables

- Positioning of grinding heads on grinding machines - Small numerically controlled systems (EMCO) - Tape and floppy disk drives, ...;

- Printers - **Regulation:**

- Valve opening position control;

A linear motor is essentially an electric motor that "has been

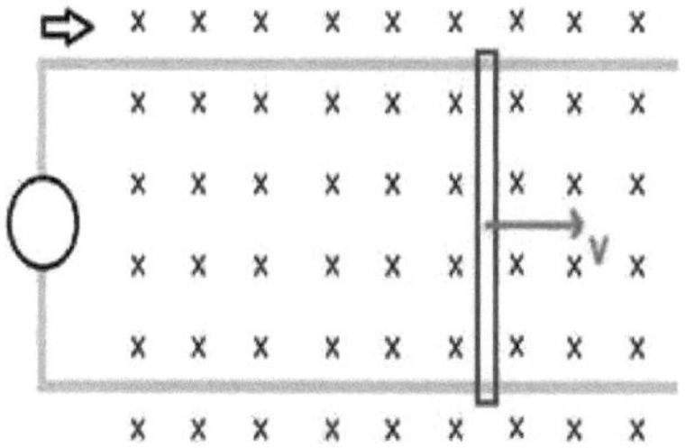

In this way, instead of producing a torque (rotation), it produces a linear force

along its length by installing an electromagnetic field of
movement. It therefore requires far fewer adaptations than conventional approaches, where linear movement is achieved by coupling a rotary motor to a ballscrew or gear rack. There are therefore fewer moving parts and less inertia and backlash. As a result, the linear motor is the ideal choice when speed and precision really matter.

- Operating principle :

Experience of the Laplace rail :

The Laplace rail is a simple linear motor, and is the fundamental experiment illustrating the operation of an electric motor. A cylindrical metal rod, placed in contact with two horizontal electrically conductive rails and closing an electric circuit through which a direct current flows, the whole being placed in a uniform vertical magnetic field, is then subjected to a Laplace force.

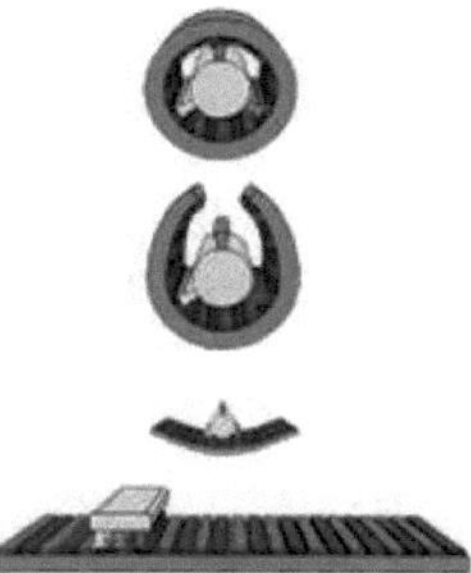

If is the length of the rod, the value of the magnetic field, and the value of the current, the Laplace force is :

The rod accelerates and, as it moves, generates a counter-electromotive force proportional to its speed (Lenz-Faraday law, with the rod cutting off the magnetic flux), with an induced current in the opposite direction.

The counter-electromotive force gradually compensates for the electromotive force, and the intensity tends towards zero in the absence of opposing mechanical forces. The rod therefore reaches a limit speed.

- Motorisation:

Today's engines use alternating current, and a specially designed track.

> Two principles can be applied:

- if there are two magnetic fluxes, 1 in the channel and 1 in the part
mobile, it is a *synchronous* motor
- if the magnetic flux is only generated at one point, for example in the receiver, then it may be a simple metal conductor (copper, aluminium, etc.).
the track, and the moving part is electro-magnetically passive, the motor is said to be *asynchronous4*.

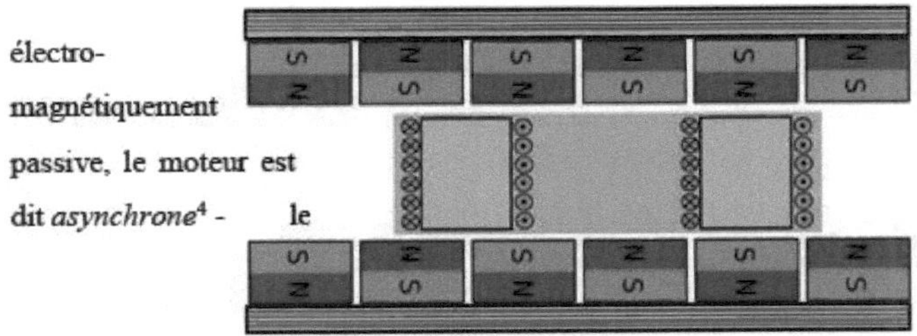

In the case of a synchronous linear motor, the track can be made up of permanent magnets, with an alternating sequence of north and south poles. Coils attached to the tractor, supplied with alternating current, and suitably phased, enable the speed to be synchronised (depending on the frequency of the current and the spacing of the magnets). However, the track can be marked out with electromagnets supplied with alternating current, and permanent magnets placed in the moving part, as in the Maglev or Transrapid, allowing levitation. The speed of the train is then synchronised with the magnetic wave of the loops placed in the track.

In the case of a linear asynchronous motor, the magnetic field generated in the track is alternating, with at least three phases and a wave moving at a speed v1. In the moving part, the receiver may be a simple conductive plate, or parallel strands (perpendicular to the track) connected at the ends, like the *squirrel cage* of a flattened asynchronous rotary motor. Eddy currents (if the plate or rod is long), or induced currents (if the bars are long), opposing the variation in magnetic flux (Lenz's law), set the tractor in motion, which reaches a speed v2 almost equal to v1 (the difference being the *slip*). Conversely, the magnetic field can also be generated in the moving part, and the secondary part (the rail) can be a simple conductive plate where the eddy currents are induced.

One of the main advantages of the linear motor is its strength at low speeds, its precision, and less wear and tear (fewer contacts, because you get a force directly and not a torque).

-Linear motor type :

For each type of linear motor, there is a corresponding type of rotary motor. This gives the same classification as for rotary motors. But linear motors can also be classified according to their geometry.

-Classification of linear motors by geometry :

There are two main types of linear motor: the flat geometry linear motor and the tubular geometry linear motor. They can be further divided into two parts according to the geometry of the primary: long or short. Flat geometry linear motors can be further subdivided according to the number of primaries: double-primary and single-primary.

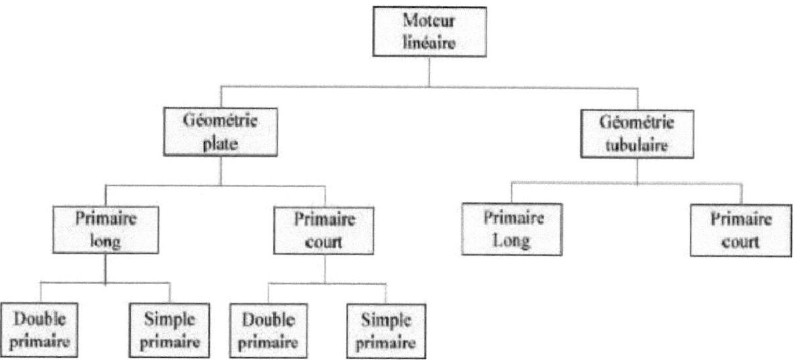

-Classification of linear motors according to their magnetic circuit :

There is another classification based on the engine's operating principle.

This is shown in the figure. Electromagnetic linear motors are the most widely used and can be divided into three parts: inductive, synchronous and direct current linear motors.

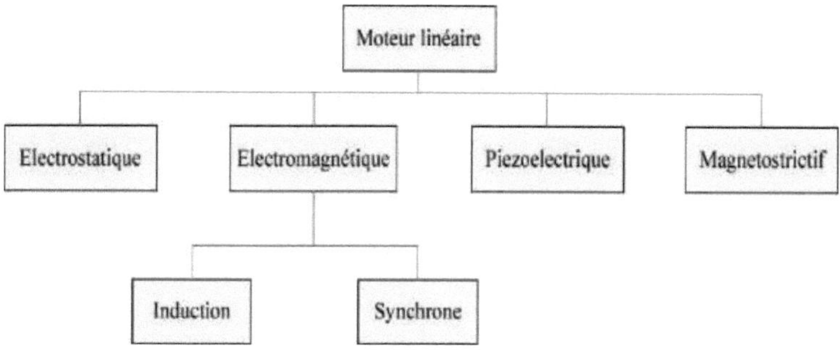

❖ Linear motor application :

Hard disk, CD-ROM and DVD drive arm :

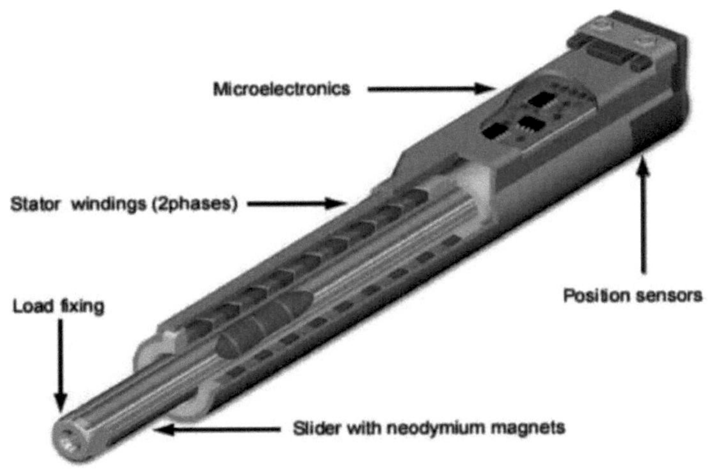

Linear motors are very popular as actuators for the hard disk drive arms of our microcomputers, because of their compactness and the speed and precision of their movements.

❖ **The Maglev motor train :**

No more wheels, so no more noise, apart from that of the inverter (which converts the direct current into a higher alternating current). The electromagnets lift the train 6mm above the guide surface to propel it.

The first commercial linear motor train (nicknamed "Linino") will travel 7km from Fujigaoka station to the North Gate of the site. Similar systems have already been tried out for transport, but this will be the first commercial line to use the "Maglev" (magnetic levitation) linear motor train under normal conduction.

The construction cost of a Linimo train is slightly higher than that of a conventional train, but the operating costs can be offset, firstly because it is fully automatic, and also because there is virtually no natural wear and tear on the vital components, as the trainsets are not in contact with the guideway surface.

The Maglev transport system will one day be the ideal candidate to replace long and medium-distance trains.

The three carriages of a Linimo trainset seat 244 people, rising to 370 at peak times.

Maximum speed: 100 km/h. Trains will travel at 6-minute intervals during morning and evening rush hours, and at 10-minute intervals during normal hours.

On 31 December 2002, Chinese Prime Minister Zhu Rongji and German Chancellor Gerhard Schroder inaugurated the magnetic levitation and linear motor train link between the economic centre of Shanghai and Pudong International Airport.

The 30 km journey currently takes 8 minutes (average speed = 225 km/h). The maximum operating speed has been set at 420 km/h.

This German-designed train (Siemens-ThyssenKrupp consortium), powered by a linear motor, comprises three carriages and is magnetically levitated using conventional magnets and coils.

By contrast, the Japanese Maglev and Swissmetro prototypes use the Meissner effect for their lift (the Meissner effect is the repulsion between magnets and superconductors due to the perfectly diamagnetic behaviour of superconductors).

CHAPTER 8

Permanent magnet synchronous machine

1. Introduction

Following the Danish chemist 0rsted's discovery of the phenomenon of the link between electricity and magnetism, electromagnetism, Ampere's theorem and Biot and Savart's law, the English physicist Michael Faraday built two devices to produce what he called "electromagnetic rotation": the continuous circular movement of a magnetic force around a wire, in effect demonstrating the first electric motor.

In our business, permanent magnet synchronous motors are finding new applications, particularly in EC motors (electronically commutated motors). This technology is being used more and more, particularly in ventilation and extraction systems. These motors are used for speed variation, for example to control condensation pressure in refrigeration systems.

In this work we present the permanent magnet synchronous machine, its construction, general principle, types of MSAP, the use of MSAP and the advantages and disadvantages.

2. Definition of permanent magnet synchronous machine :

Electrical machines are electromechanical devices based on electromagnetism that convert electrical energy into work or mechanical energy, for example. This process is reversible and can be used to produce electricity:

- Electrical machines that produce electrical energy from mechanical energy are commonly referred to as generators, dynamos or alternators, depending on the technology used.
- Electrical machines that produce mechanical energy from electrical energy electrical energy are commonly referred to as motors.

For example, three-phase asynchronous cage motors are no longer used solely for coupling, and thanks to power electronics (variable frequency drives) their field of application has changed considerably.

However, as all these electrical machines are reversible and can behave either as a "motor" or as a "generator" in the four quadrants of the torque-speed planeN

1,1,2,3, the motor/generator distinction is made "communally" in relation to the final use of the machine.

Rotary motors produce energy corresponding to the product of torque and angular displacement (rotation), whereas linear motors produce energy corresponding to the product of force and linear displacement.

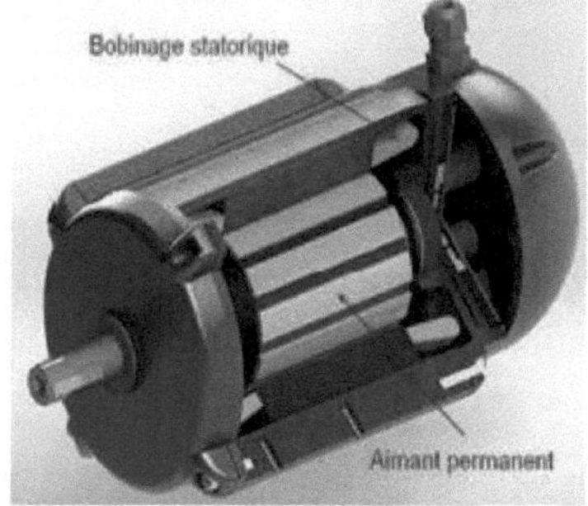

Parmanent magnet synchronous machines are energy convertersmechanical into energy electric machine (alternator) or conversely (synchronous motor),is an alternating-current electric machine with a stator,the rotor of which rotates at exactly the synchronous speed of the statically rotating chumps given by

$$N_s = \frac{60 f_s}{p} = N$$

fs stator current frequency

Ns=mechanical rotor speed[rd/s]

P: number of pole pairs

The MSAP is equipped with magnets attached to the rotor (there is also another rotor variant, the MS with an excitation circuit: excitation coil carried by DC current j or If).

Permanent magnet motors can accept high overload currents for rapid start-up. Combined with electronic variable speed drives, they are used in certain lift drive applications where a certain degree of compactness and rapid acceleration are required (high-rise buildings, for example). In this case, the excitation is created by permanent magnets. The instantaneous torque (in any magnet machine configuration) is the sum of three elementary torques: the reluctant torque, the hybrid torque and the detent torque.

3. Construction of the MSAP :

The MSAP consists of :

Stator: this is the fixed part and comprises a 3-phase winding through which flows an alternating current/voltage system with a frequency of fs=50/60 HZ. This coil creates a magnetic chump rotating statorically with a pulsation of

$W_s = w = p\Omega_s$

The conductors of a stator winding are laminated to reduce magnetic losses (disc of magnetic sheets isolated from each other).

Rotor :

this is the rotating part, it carries the excitation current and is a permanent magnet with the same number of poles as the stator, producing a rotating magnetic rotor chump (the rotor is polar and drives the stator chump).

4. Operating principle :

Alternator: the rotor is driven (single-phase coil with 2 poles and a DC winding). The stator, a 3-phase coil with 2 poles, will be subjected to a rotor shock (the stator is the induced EMF snow).

If the stator coil delivers a balanced 3-phase load, it delivers a balanced 3-phase

system with pulsation $W_s=p\Omega$

Moter : the stator is supplied with SCTE power ws to create a rotating chump at Q=ws/p . the rotoriqe winding is wound in dc still at standstill and consequently the rotor is subjected to induced EMF torque .

5. Types of MSAP :

There are two types of MSAp :

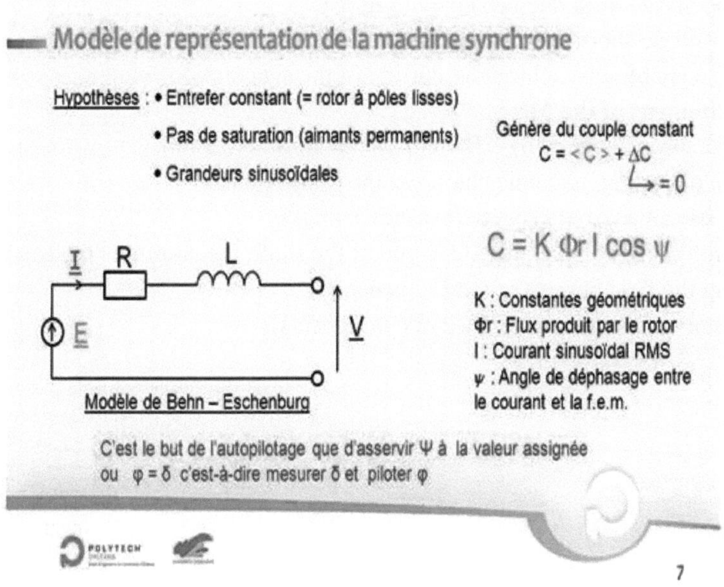

Machine with smooth poles: the field winding is distributed in slots in a similar way to the armature, in this case the reluctance of the magnetic circuit between stator and rotor is practically constant and independent of the rotor position (constant air gap).

These machines are robust and can withstand high rotation speeds, high power machines (50,000 KVA) (turbo alternator).

2-machine with salient poles: where the field winding is concentrated on a spoke, in addition to the excitation winding. The MS rotor has a short-circuit cage similar to that of the Mas sit "amortisseur".

Rotor à pôles saillants

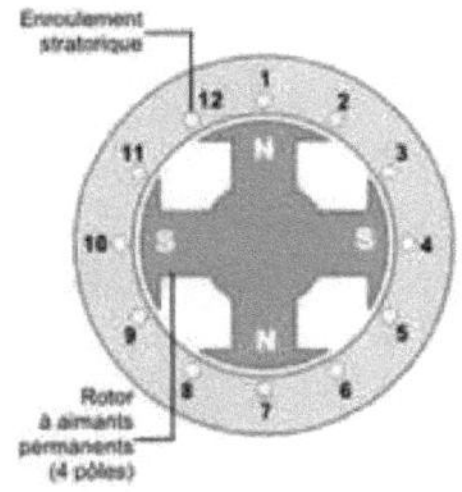

Rotor à pôles lisses

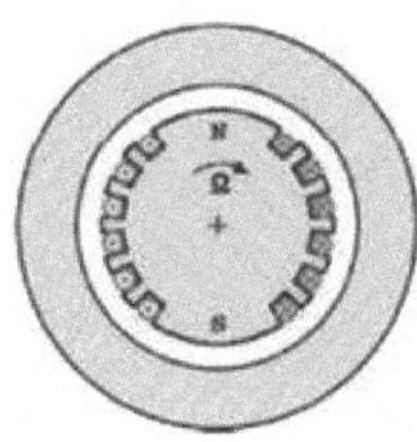

Role of amators: as its name suggests, damping :

1- Happens at any sudden change of regime
2- Two-phase short-circuit, one phase open
3- It also enables asynchronous starting of the synchronous motor.

6. Excitement of the MS :

The inductor is powered by a DC current and is an electromagnet whose role is to create a rotating magnetic chump as the rotor rotates.

We can use an auxiliary source (brushes + rings)

Ulitisation of a DC generator at the end of the shaft, the current is adjusted by adjusting the field current of the DC generator.

Use of static excitation (no collectors, no fencers).

Power balance
Bilan de puissance

Considérons à nouveau le circuit équivalent de référence des machines à pôles lisses.

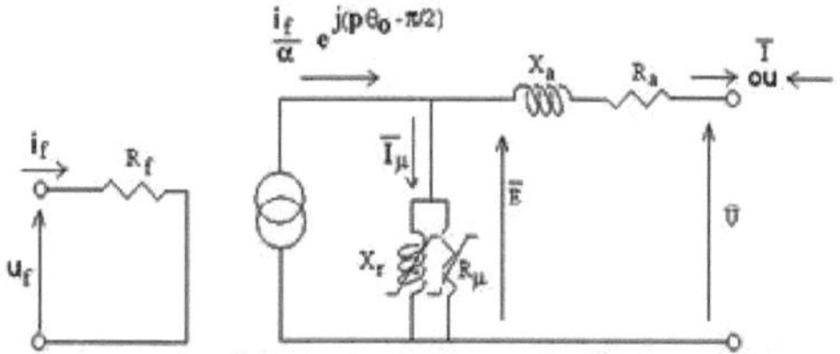

Les grandeurs de ce circuit équivalent ont une signification physique.

3 R_a I^2 correspondent bien aux pertes par effet Joule.

3 E^2/R_μ ou $E_L{}^2/R_\mu$ correspondent bien aux pertes magnétiques du stator (que dire du rotor ?).

R_f $I_f{}^2$ correspond bien aux pertes d 'excitation.

La puissance convertie en puissance mécanique est donc clairement le triple de la puissance de la source de courant !

ELEC2753 - 2012 - Université catholique de Louvain

Let's look again at the equivalent reference circuit for machines with smooth poles .

The quantities in this equivalent circuit have a physical meaning.

3 R_a I^2 correspond to Joule effect losses.

3 E /$R^2{}_A$ or $E_L{}^2$ /R^ correspond to the magnetic losses of the stator (what about the rotor?).

R_f If^2 corresponds to the excitation losses.

The power converted into mechanical power is clearly three times the power of the current source I

ELEC2"53 - 2012 - Catholic University of Leuven

7. The use of MS :

The synchronous machine is often used as a generator. It is then called an "alternator". With the exception of low-power generators, these machines are generally three-phase. To produce electricity, power stations use alternators whose power can be in the region of 1,500 MW.

As the name suggests, the speed of rotation of these machines is always proportional to the frequency of the currents flowing through them. This type of machine can be used to increase the power factor of an installation. This is known as a "synchronous compensator".

Synchronous machines are also used in traction systems (such as the TGV); in this case, they are often combined with current inverters, which enable the motor torque to be controlled with a minimum of current. This is known as autopilot control (servo-control of stator currents in relation to rotor position).

MSAPs are used for small power ratings (<10KW). The rotor is a parmanent magnet and has no commutator or brushes.

they are used in certain lift motorisation applications where a certain degree of compactness and rapid acceleration are required (high-rise buildings, for example).

8. Advantages and disadvantages

The synchronous machine with permanent magnets on the surface seems to be the best choice for the wheel motor. These machines do indeed have significant advantages:

High torque/mass and power/mass ratios.

Very good performance.

Less wear and tear and lower maintenance costs (no brushes or carbon brushes).

9. However, they do have certain drawbacks:

- High cost (due to the price of magnets).
- Problems with magnet temperature resistance (250°C for samarium-cobalt)
- Risk of irreversible demagnetisation of the magnets due to armature reaction.
- Difficult to deflux and complex control electronics (position sensor required).
- Impossible to regulate arousal.
- To reach high speeds, it is necessary to increase the stator current in order to demagnetise the machine. This will inevitably lead to an increase in stator losses due to the Joule effect.
- The fact that this flow is not regulated means that it cannot be controlled flexibly over a very wide speed range.

10. Conclusion

This work is a presentation on permanent magnet synchronous machines, which are generally three-phase machines. The rotor, often called "pole wheel", is supplied by a direct current source or equipped with permanent magnets. They play a very important role in the industrial field because they are :

High torque/mass and power/mass ratios.

Very good performance.

Less wear and tear and lower maintenance costs

But they also have certain disadvantages, such as :

High cost.

Problem with magnet temperature resistance.

control (position sensor required).

Impossible to regulate arousal.

To reach high speeds, it is necessary to increase the stator current in order to demagnetise the machine.

Bibliography

1. Quadrants II or IV of the torque-speed plane (known as the "four quadrants"), shown in the article "Quadrant (mathematics)", with speed on the ordinate and torque on the abscissa. Like all electrical machines - which are by nature reversible - a synchronous machine switches seamlessly between "motor" and "generator" operation simply by reversing the sign of the torque (driven or driving load, for example during acceleration or braking phases) or the sign of the speed (reversal of the direction of rotation).

2. BTS Electrotechnique (second year) - Machine a courant direct - Quadrants de fonctionnement [archive], site physique.vije.net, consulted on 8 August 2012.

3. a[et] [b] Robert Chauprade, Francis Milsant, *Commande electronique des moteurs a courant alternatif - A I'usage de I'enseignement superieur, ecoles d'ingenieurs, facultes, CNAM,* Paris, ed. Eyrolles, coll. " Ingenieurs EEA ", 1980, 200 p., p. 86-92.

4. In quadrants I or III of the torque-speed plane defined in the note above.

5. Description of a synchronous motor [archive], on sitelec.org, 7 September 2001, accessed 28 March 2012.

6. a[et b] Ilarion Pavel, p. 18-28.

7. a[et b] **(en)** P. Zimmermann, "Electronically Commutated D.C. Feed Drives for Machines Tools", Robert Bosch GmbH - Geschaftsbereich Industrieaurustung, Erbach, Germany, p. 69-86, in *Proceding of PCI Motorcon*, September 1982, p. 78-81.

8. a b et c " Dynamic stability of industrial electrical networks " [archive] (consulted on 18 December 2012) [PDF].

9. Diagram taken from *Grundlagen der Hochspannungs- und Energieubertragungstechnik*, TU Munich, p. 246.

10. Mikhail Kostenko and Ludvik Piotrovski, Electrical Machines, *t. II,* Alternating Current Machines, *Moscow Publishing House (MIR), 1969; 3rd edition, 1979, 766p.*

11. Ilarion Pavel, "Nikola Tesla's invention of the synchronous motor" *[archive] [PDF],*

12. on bibnum.education.fr [archive],*bibnum [archive], January 2013.*

Printed by Books on Demand GmbH, Norderstedt / Germany